What's it like to be an...
AIRLINE PILOT

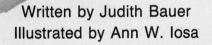

Written by Judith Bauer
Illustrated by Ann W. Iosa

Troll Associates

Special Consultant: Captain Thomas Van Note, *Pilot, Pan American Airlines.*

Library of Congress Cataloging-in-Publication Data

What's it like to be an airline pilot / by Judith Bauer;
illustrated by Ann W. Iosa.
    p.   cm.—(Young careers)
    Summary: Describes the work of an airline pilot as he goes about
getting his plane into the air, completing the scheduled flight, and
landing it.
    ISBN 0-8167-1791-5 (lib. bdg.)   ISBN 0-8167-1792-3 (pbk.)
    1. Jet transportation—Piloting—Vocational guidance—Juvenile
literature. [1. Air pilots. 2. Occupations.] I. Iosa, Ann W., ill.
II. Title. III. Series.
TL710.S645 1990
629.13'023—dc20   89-34397

What's it like to be an...

AIRLINE PILOT

Captain Ross Turner is an airline pilot. He is
flying a jumbo jet to California today. It is a
special flight for Captain Turner—his son and
daughter will be passengers aboard his jet.

Captain Turner takes his children to the
departure gate. They will board Flight 504 to
Los Angeles in one hour.

"See you in California, Dad," Kate says.

"Be good," Captain Turner tells them. "I have to go to work now."

"Yes, sir!" Tim says with a salute.

Kate and Tim walk over to a large window.
They can see their plane on the ground below.

Captain Turner goes to the flight dispatcher's office to get information about his flight.

"Good morning, Captain Turner," the dispatcher says. "Here's the weather report for Flight 504."

The captain carefully reads the up-to-the-minute report. He finds out about weather conditions all across the country, where his plane will fly.

"Just what I ordered!" Captain Turner says with a smile. "Sunny skies all the way!"

"Glad to hear it!" says the copilot for Flight 504. He is the pilot's right-hand man. "Kate and Tim should have a nice, smooth flight to L.A.!"

He looks over the navigation charts with Captain Turner.

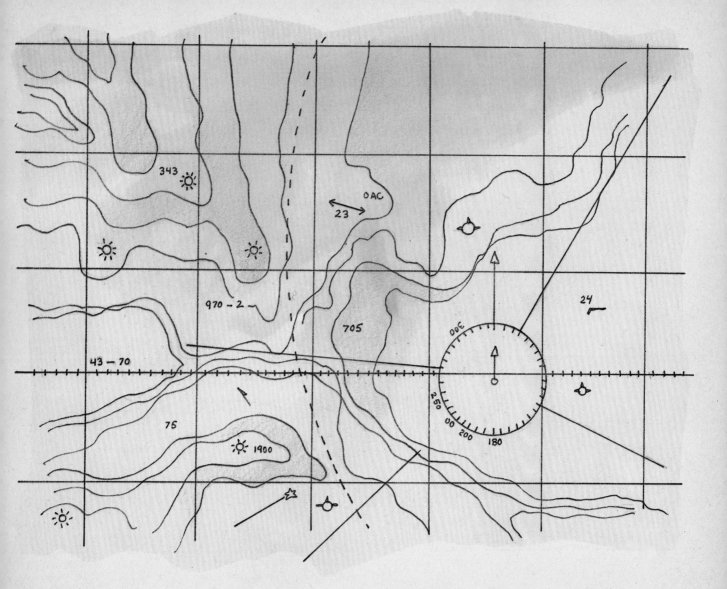

Navigation charts are used to make a flight
plan. The flight plan shows the best path for the
jet to take. It tells how high the plane should
fly. It tells how much fuel the jet needs.

Meanwhile, the flight engineer checks out the plane. He is the third member of the flight crew. It is his job to make sure that everything is working before the plane takes off. First, he watches the jet being fueled. Then, he inspects the tires, the wings, and the outside of the plane. He checks to see that the plane is in fine shape.

Tim and Kate watch from the terminal. They see the cargo loader carrying baggage from the terminal to the plane.

"Can you find our suitcases, Tim?" Kate asks. "Look for the ones with stickers all over them!" Tim and Kate have stickers that their father brings home from all over the world.

"Look, Kate!" says Tim, pointing to a truckload of aluminum-covered trays. "There's our lunch! I wonder what we're having today?"

Inside the terminal, Captain Turner meets with his crew. He explains the flight plan. He answers any questions they have.

Restaurant

Snack Bar

Ticket Desks

Departures Area

Baggage Claims

Customs

Arrivals Area

When the crew boards the plane, the pilot, copilot, and flight engineer will go into the cockpit. The flight attendants will go into the cabin.

In the cockpit, the pilot and copilot inspect the jet's instruments and computers. The flight engineer checks the fuel, oil and electrical systems.

Pilot

Copilot

Flight Engineer

In the cabin, the flight attendants greet the passengers.

"Welcome to Flight 504," they say, as they help the passengers to their seats. The crew makes sure everyone is ready for take-off.

Tim and Kate buckle up their seat belts. They are excited about flying.

A jet cannot go backward all by itself. So a machine called a tug gently pushes the jet away from the gate. Then Captain Turner starts up the engines. He radios the control tower at the airport. Flight 504 is ready to go!

The ground controller tells the pilot which taxiway to use. The taxiway leads from the boarding area to the runway.

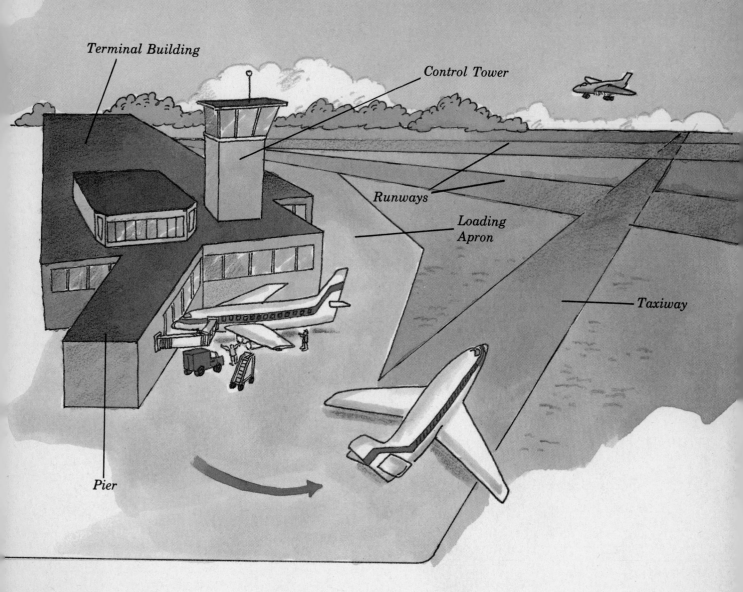

Terminal Building

Control Tower

Runways

Loading Apron

Taxiway

Pier

Captain Turner taxis the big jet to its runway. At this time, the flight attendants explain the safety rules.

"Please keep your seat belts fastened for take-off," they say.

19

Now the pilot radios the control tower again. An air-traffic controller sits at the top of the tower. She supervises take-offs and landings. When she is sure there is no traffic along the flight path, she gives the pilot the okay for take-off.

The huge jet engines begin to roar. Captain Turner steers the plane down the runway. It picks up more and more speed. As the pilot pulls back on the yoke, or steering control, the jet lifts up into the air.

Pilot

Copilot

Flight
Instruments

Engine
Instruments

Weather
Radar

Throttles

The pilot and copilot work together to put the
plane on course. A special computer guides them
along their flight path.

When the jet is safely on its way, the copilot takes over the controls. Captain Turner speaks to the passengers over the plane's loudspeaker.

"Welcome to Flight 504. This is your captain speaking. We are cruising at 600 miles an hour today. Our altitude is 30,000 feet. Weather conditions permitting, we should arrive at Los Angeles International Airport at 1:00 P.M. Enjoy your trip."

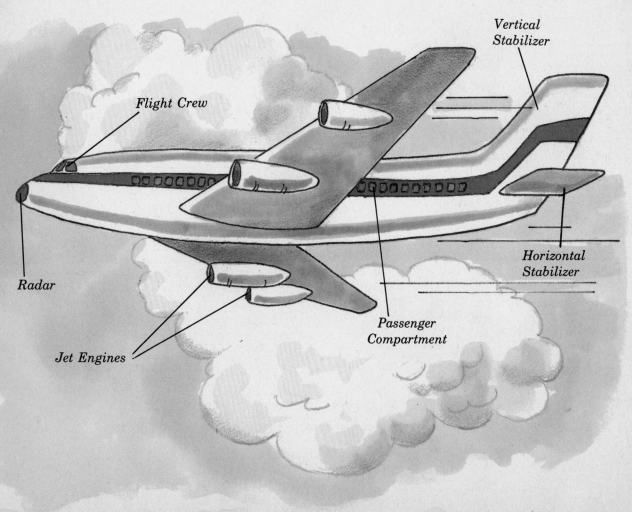

Vertical Stabilizer

Flight Crew

Radar

Horizontal Stabilizer

Passenger Compartment

Jet Engines

The cockpit crew carefully watches the plane's instruments during the flight. The radar shows weather conditions. If the plane approaches bad weather, the pilot will change course.

The cabin crew is busy, too. They serve the passengers food and make sure everyone is comfortable.

"Would you like fried chicken or a cheese omelet for lunch?" the flight attendant asks Kate and Tim.

"Fried chicken!" they both reply.

When lunch arrives, there is a surprise for Kate and Tim. On top of each tray is a pair of wings.

"Just like Dad's," Kate says.

The children gaze out at the puffy white clouds below.

"It looks like white cotton candy," Tim says.

"Or a big, soft quilt," Kate adds.

Soon the plane is near the Los Angeles airport. Captain Turner radios the control tower to find out which runway to use for landing.

In bad weather, Captain Turner uses a
special computer to help land the plane. The air-
traffic controller tells him what path to follow to
the ground.

But today the sky is clear. Captain Turner
can see the long, gray ribbon of runway below.
He will bring the plane down by sight.

The jet hits the runway with a soft bump. It's a smooth landing! Captain Turner slows down the plane and turns off the runway onto a taxiway. Finally, he stops the plane by the terminal building.

"Nice job," the pilot tells his crew. They shut down their instruments.

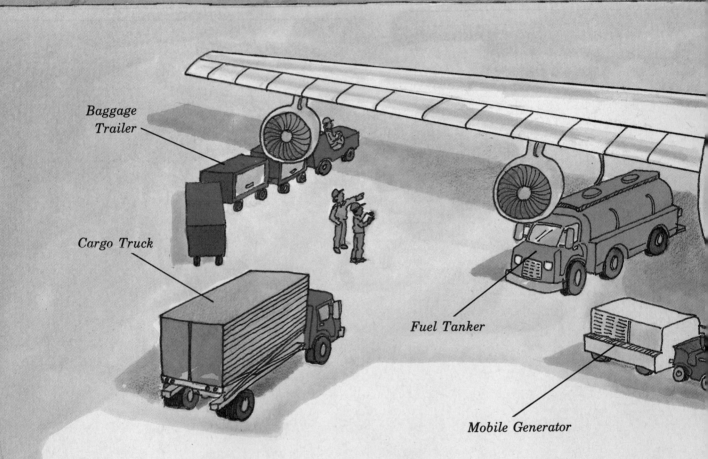

Baggage Trailer

Cargo Truck

Fuel Tanker

Mobile Generator

The passengers leave the plane. The cargo handlers unload the baggage. The mechanics check out the plane for its next flight.

Waste Tanker

Mechanic

Captain Turner puts on his blue captain's
hat. He joins Tim and Kate inside the terminal.

Inside the terminal, the children hear, "Kate! Tim! Over here!" Their grandparents are waiting. Tim and Kate rush to meet them. It is the end of an exciting flight, but the beginning of a wonderful vacation.